THE SEED OF THE TOC-TOC BIRDS

FRANCIS FLAGG

Édition et mise en page : Culturea (Hérault, 34)
Retrouvez notre catalogue sur http://culturea.fr/
Contact : infos@culturea.fr
Imprimé en Allemagne par Books on Demand Gmbh
Design typographique et layout : Derek Murphy et Reedsy
ISBN :9791041985586
Date de parution : février 2024
Ce livre :
— a été composé avec une police open source libre de droits dessinée sur Open Fondry https://open-foundry.com/

— permet de replanter un arbre. SHS Éditions soutient Plantons Pour l'Avenir, fonds de dotation pour le reboisement des forêts françaises
400 projets de reboisement de forêts soutenus depuis 2014 à travers la France https://www.plantonspourlavenir.fr/

TALBOT had been working that day, far up in the Catalinas, looking over some mining prospects for his company, and was returning to the Mountain View Hotel in Oracle when, from the mouth of an abandoned shaft some distance back of that town, he saw a strange object emerge.

"Hello," he said to Manuel, his young Mexican assistant, "what the devil can that be?"

Manuel crossed himself swiftly.

"Dios!" he exclaimed, "but it is a queer bird, señor."

Queer, it certainly was, and of a species Talbot had never before laid eyes on. The bird stood on the crumbling rim of the mining shaft and regarded him with golden eyes. Its body was as large as that of a buzzard, and its head had a flat, reptilian look, unpleasant to see. Nor was that the only odd thing. The feathers glittered metallically, like blued copper, and a streak of glistening silver outlined both wings.

Marveling greatly, and deciding that the bird must be some rare kind escaped from a zoo, or a stray from tropical lands much further south, Talbot advanced cautiously, but the bird viewed his approach with unconcern. Ten feet from it he stopped uneasily. The strange fowl's intent look, its utter immobility, somewhat disconcerted him.

"Look out, señor," warned Manuel.

Involuntarily, Talbot stepped back. If he had possessed a rifle he would have shot the bird, but neither Manuel nor himself was armed. Suddenly—he had looked away for a moment—the bird was gone. Clutching a short miner's pick-ax, and a little ashamed of his momentary timidity, he strode to the edge of the abandoned shaft and peered down. There was nothing to see; only rotting joists of wood, crumbling earth for a few feet, and then darkness.

HE pondered for a moment. This was the old Wiley claim. He knew it well. The shaft went down for over two hundred feet, and there were several lateral workings, one of which tunneled back into the hills for a considerable distance. The mine had been a bonanza back in the days when Oracle boomed, but the last ore had been taken out in 1905, and for twenty-seven years it had lain deserted. Manuel came up beside him and leaned over.

"What is that?" he questioned.

Talbot heard it himself, a faint rumbling sound, like the rhythmic throb of machinery. Mystified, he gazed blankly at Manuel. Of course it was impossible. What could functioning machinery be doing at the bottom of an abandoned hole in the ground? And where there were no signs of human activity to account for the phenomenon? A more forsaken looking place it would be hard to imagine. Not that the surrounding country wasn't ruggedly beautiful and grand; the hills were covered with live-oak, yucca grass, chulla, manzanita, and starred with the white blossoms of wild thistle. But this locality was remote from human habitation, and lonely.

Could it be, Talbot wondered, the strange bird making that noise? Or perhaps some animal? The noise sounded like nothing any creature, furred or feathered, could make, but, of course, that must be the explanation. However, it would be dark within the hour, with Oracle still two miles distant, so he turned reluctantly away, Manuel thwacking the burros from the grazing they had found. But that was not to be the end of the odd experience. Just before the trail swung over the next rise, Talbot glanced back. There, perching on the rim of the abandoned mining shaft, were not one but two of the strange birds. As if cognizant of his backward glance, they napped their gleaming, metallic wings, although they did not rise, and gave voice to what could only be their natural harsh cries, measured and, somehow, sinister.

"Toc-toc, toc-toc."

Talbot went to bed determined to investigate the old Wiley claim the next day, but in the morning an urgent telegram called him and Manuel to

Phoenix, and so the matter was necessarily postponed. Moreover, on mature reflection, he decided that there was nothing much to investigate. The days went by, the matter slipped his mind, and he had almost forgotten the incident.

IT was an Indian who first brought news of the jungle to Oracle. His name was John Redpath and he wasn't the average person's idea of an Indian at all. He wore store clothes and a wide-brimmed hat, and spoke English with the colloquial ease of one whose native language it was. It was ten o'clock in the morning, the hour when people gathered at the local store and post-office to gossip and get their mail, when he came driving into town in his Ford, his terrified wife and three children crowded into the back seat.

"What's the matter, John?" asked Silby, the constable.

"Matter?" said Redpath. "I'll tell you what's the matter."

He held the attention of the crowd which now began flocking around him. "You know me, Silby; I'm not easily frightened; but what's happened at my place has me scared stiff."

He pulled out a handkerchief and mopped his brow.

"When we went to bed last night, everything looked as usual; but this morning...."

He paused.

"Something over night had grown up in my pasture. Don't ask me what it is. The whole hillside was filled with it. I went to the pasture to milk my goats—that's some distance from the house and over a rise; you know how rugged my land is—and there was the stuff, acres of it, twenty, thirty feet tall, like—like nothing I had ever seen before. And Silby"—his voice was suddenly low—"I could see it growing."

AT this remarkable statement, everyone in sound of his voice gaped with astonishment. Had it been any other Indian they would have said he was drunk—but not John Redpath. He didn't drink.

"Growing?" echoed Silby stupidly.

"Yes. The damn stuff was growing. But it wasn't that which stampeded me out of there. It was the globe."

"The globe!" said Silby, more mystified than ever.

"It was floating over the growing stuff, like a black balloon. Just over my place the balloon began to sift down a shower of pebbles. Like beans, they were; seeds, rather; for when they hit the ground they started to sprout."

"Sprout?" The constable was capable of nothing more than an echo.

"I'm telling you the truth," continued Redpath. "Incredibly fast. I had barely time to crank up the car and get out of there. I never would have done it if the strange growth hadn't left the way clear from the garage to the road. Silby, I had the devil of a time getting the wife and kids out of the house. When I looked back after going a quarter of a mile the house had disappeared under a tangled mass."

There was no time for anyone to question John Redpath further. Even as he finished speaking a large automobile dashed up and out tumbled a well-dressed and portly red-faced stranger.

"What the devil's the matter with the road above here? Funniest thing I ever saw. The road to Mount Lemmon's blocked. My family," he said inconsequentially, "is at Mount Lemmon for the summer and I want to get through to them."

Blocked! The crowd stared at him wonderingly. John Redpath threw in his clutch. "So long," he said. "I've a brother in Tucson, and I'm going to his place until this blows over."

As he left Oracle, John Redpath noticed several dark globes drifting down on it from the hills.

THE first inkling the outside world had of the terrible tragedy that was happening at Oracle came over the phone to Tucson while John Redpath was still en route to that city.

"Hello, hello! Is this the police station? Silby speaking. Silby, town constable at Oracle. For God's sake, send us help! We're being attacked. Yes, attacked from the air. By strange aircraft, round globes, discharging—oh, I don't know what it is; only it grows when it hits the earth. Yes, grows. Oracle is hemmed in. And there are the birds—b-i-r-d-s, birds— —"

There was a stifled cry, the voice suddenly ceased, and the wire went dead.

"My God!" said the chief of police of Tucson, "somebody's raving." He lost no time in communicating with the sheriff's office and sending out his men. They soon returned, white-faced and shaken.

"Chief," said the officer in charge of the party, "you know where the road to Oracle switches off the main highway? Well, it's impassable, covered with stuff a hundred feet high."

The chief stared. "Are you crazy?"

"No. Listen. It's the queerest growth you ever saw. Not like vegetation at all. More like twisted metal...."

BUT now the city began to seethe with excitement. Farmers and their families flocked in from the Seep Springs district, and from Jayhnes, telling weird tales of drifting globes and encroaching jungle. The Southern Pacific announced that traffic northward was disrupted. Extras appeared on the streets with shrieking headlines. Everything was in confusion.

A flyer from the local airport flew over Oracle and announced on his return that he could see no signs of the town, that its immediate vicinity was buried under an incredibly tall and tangled mass of vegetation. "From the air it looks like giant stalks of spaghetti, twisted, fantastic," was his description. He went on to say that he noticed quite a few drifting globes and large birds with black, glistening wings, but these offered no hindrance to his flight.

Now the wires hummed with the startling news. All the world was informed of the tragedy. The great cities of the nation stood aghast. An aroused Washington dispatched orders for the aerial forces of the country to proceed to Arizona without delay. The governor of Arizona mobilized the state militia. All border patrol officers proceeded to the area affected. And yet in the face of what was happening they were powerless to do a thing.

At two o'clock of the day following the wiping out of Oracle, the first black globes approached Tucson. They floated down from the north, skirting the granite ridges and foothills of the Catalinas, and were met with a withering hail of lead from anti-aircraft guns, and burst, scattering wide their contents. When some three hours later the first squadron of the air fleet came to earth on the landing field a few miles south of the city, the northern environs of Tucson, all the area the other side of Speedway, and running east and west as far as the eye could see, was a monstrous jungle a hundred or more feet tall—and still growing.

TERRIFIED residents fled before the uncanny invasion. People congested the streets. Thousands fled from the city in automobiles, and thousands of others thronged the railroad station and bus-line offices seeking for transportation. Rumors ran from lip to lip that Russia was attacking the United States with a newly invented and deadly method of warfare; that it wasn't Russia but Japan, China, England, Germany, a coalition of European and Asiatic powers.

Frantically, the city officials wired railroad companies to send in emergency trains. The mayor appealed to the citizens to be quiet and orderly, not to give way to panic, that everything was being done to insure their safety. Hastily deputized bodies of men were set to patrolling streets and guarding property. Later, martial law was established. The south side of Speedway rapidly assumed the appearance of an armed camp. At the landing field Flight Commander Burns refueled his ships and interviewed the flyer who had flown over Oracle. That worthy shook his head.

"You're going out to fight, Commander," he said, "but God knows what. So far we have been unable to detect any human agency back of those globes. They just drift in, irrespective of how the wind is blowing. So far our only defense has been to shoot them down, but that does little good; it only helps to broadcast their seed. Then, too, the globes shot down have never been examined. Why? Because where they hit a jungle springs up. Sometimes they burst of their own accord. One or two of them got by us in the darkness last night, despite our searchlights, and overwhelmed a company of National Guards."

The flight commander was puzzled.

"Look here," he said, "those globes don't just materialize out of thin air. There must be a base from which they operate. Undoubtedly an enemy is lurking in those mountains." He got up decisively. "If it is humanly possible to locate and destroy that enemy, we shall do it."

FLYING in perfect formation, the bombing squadron clove the air. Looking down, the observers could see the gigantic and mysterious jungle which covered many square miles of country. Like sinuous coils of spaghetti, it looked, and also curiously like vast up-pointed girders of steel and iron. The rays of the late afternoon sun glinted on this jungle and threw back spears of intense light. Over the iron ridges of the Catalinas the fleet swept at an elevation of several thousand feet. Westward, numerous huge globes could be seen drifting south. The commander signaled a half dozen of his ships to pursue and shoot them down.

In the mountains themselves, there was surprisingly little of the uncanny vegetation. Mile after mile of billowing hills were quartered, but without anything of a suspicious nature being noted. Here and there the observers saw signs of life. Men and women waved at them from isolated homesteads and shacks. At Mount Lemmon the summer colonists appeared unharmed, but in such rugged country it was impossible to think of landing. Oracle, and for a dozen miles around its vicinity, was deserted.

Though the commander searched the landscape thoroughly with his glasses, he could detect the headquarters of no enemies; and yet the existence of the drifting globes would seem to presuppose a sizable base from which they operated. Mystified, he nevertheless subjected the Oracle area to a thorough bombing, and it was while engaged in doing so that he and his men observed a startling phenomenon.

HIGH in the heavens, seemingly out of nothing, the mysterious globes grew. The aviators stared, rubbed their eyes in amazement, doubted the truth of what they saw. Their commander recollected his own words, "Those globes don't just materialize out of thin air." But that actually seemed to be what they were doing. Out of empty space they leaped, appearing first as black spots, and in a moment swelling to their huge proportions.

One pilot made the mistake of ramming a globe, which burst, and he hurtled to earth in a shower of seed, seed which seemed to root and grow and cover his craft with a mass of foliage even as it fell. Horrified, ammunition and explosives exhausted, the amazed commander ordered his ships back to Tucson. What he had to tell caused a sensation.

"No," he said, finishing his report to the high military official who had arrived with federal forces, "I saw nothing—aside from the globes—that could possibly account for the attack. Nothing."

But none the less the attack went on. Though hundreds of planes scoured the sky, though great guns bellowed day and night and thousands of soldiers, state and federal, were under arms, still the incredible globes continued to advance, still more and more of the countryside came under the sway of the nightmarish jungle. And this losing battle was not waged without loss of human life. Sometimes bodies of artillery were cut off by globes getting beyond their lines in the darkness and hemming them in. Then they had literally to hack their way out or perish; and hundreds of them perished. One company sergeant told of a thrilling race with three globes.

"It was a close thing," he said, scratching his head, "and only a third of us made it."

FEAR gripped the hearts of the most courageous of men. It was terrifying and nerve-racking to face such an unhuman foe—weird, drifting globes and invading jungles whose very source was shrouded in mystery. Against this enemy no weapons seemed to prevail. All the paraphernalia of modern warfare was proving useless. And looking at each other with white faces—not alone in Arizona, but in New York, Chicago, Los Angeles—men asked themselves these questions, and the newspapers posed them:

"What if this thing can't be stopped?"

"What if it keeps on and on and invades every city and state?"

"It is only starting now, but what will it be like a month from now, a year?"

The whole nation awoke to a realization of its danger. The Administration at Washington solemnly addressed itself to the capitals of the world.

"If some power, jealous of the greatness of America, has perfected a new and barbarous weapon of warfare, and without due warning and declaration of hostilities has launched it against us, not only do we denounce such uncivilized procedure, but demand that such a power speak out and reveal to us and the world who our enemy is."

But the powers of the world, as one, united in disclaiming any hand in the monstrous attack being made on the United States. As for that attack, it proceeded inexorably. On the fourth day Tucson was evacuated. Then Winkleman awoke one morning to find that the drifting globes had reached the river. The town was abandoned. California mobilized citizen forces in cooperation with Nevada. The great physicist Miller was said to be frantically at work on a chemical designed to destroy the gigantic growths, specimens of which had been sent him. Such was the condition of affairs when, at Washington, Milton Baxter, the young student, told his incredible story to a still more incredulous Senate.

THE Senate had been sitting in anxious session for five days, and was little inclined to give ear to the stories of cranks. Fortunately for the world, young Baxter came of an influential family and had taken the precaution of having himself introduced by two prominent financiers, who demanded that he be heard.

"Gentlemen," he said earnestly, "contrary to current opinion, America is not being assailed by a foreign power. No! Listen to me a moment and I shall tell you what is attacking America."

He paused and held the assemblage with compelling eyes.

"But first let me explain how I know what I am going to tell you. I was in London when I read of what is occurring in Arizona. Before the wire went dead on him, didn't the unfortunate constable of Oracle say something about birds?"

The senators were silent. "Yes," said a press correspondent at length. "If I remember correctly, he said, 'And there are the birds—b-i-r-d-s, birds.'"

"Well," exclaimed Senator Huffy, "the man was pretty well excited and his words may have been misunderstood. What the devil have birds to do with those globes and jungles?"

"More than you think," replied Baxter. "Listen!" He fixed their attention with uplifted hand. "The thing I have to reveal is of such paramount importance that I must not be interrupted. You must bear with me while I go back some months and even years in time to make myself understood.

"You all remember the mysterious disappearance of Professor Reubens. Yes, I see that you do. It caused a sensation. He was the foremost scientist in the country—it would not be exaggerating too much to say in the world. His name was not as well known among the masses as that of Miller and Dean; in fact, outside of an exclusive circle it wasn't known at all, but ask any scientist about Reubens. He was a tall, dour man of sixty, with Scotch blood in his veins, and was content to teach a class in a college because of the leisure it afforded him for his own research work. That was at the University of Arizona in Tucson.

"The faculty of the college was proud to have him on its staff and provided him with a wooden building back of the campus, for a private laboratory and workshop. I understand that the Rockefeller Institute contributed funds towards Professor Reubens' experiments, but I am not certain.

"AT any rate he had a wonderfully well equipped place. I was a pupil at the University and attended his class in physics. A strong friendship grew up between us. How can I explain that friendship? I was not a particularly brilliant student, but he had few friends and perhaps my boyish admiration pleased him. I think, too, that he was lonely, heart-hungry for affection. His wife was dead, and his own boy.... But I won't go into that.

"Suffice it to say that I believe he bestowed on me some of the affection he had felt for his dead son. Indeed I am sure he did. Be that as it may, I often visited him in his laboratory and watched, fascinated, as he pored over some of his intricate apparatus. In a vague way, I knew that he was seeking to delve more deeply into the atom.

"'Before Leeuwenhoek invented the microscope,' the Professor once said, 'who ever dreamed of the life in a drop of water? What is needed now is a super-microscope to view the atom.'

"The idea thrilled me.

"'Do you believe, sir, that an instrument will ever be invented that will do that?'

"'Yes. Why not? I am working on some such device myself. Of course the whole thing has to be radically different. The present, method of deducing the atom by indirection is very unsatisfactory. We can know nothing for certain until direct observation is possible. The atomic theory that likens the atom to our solar system, with planets revolving round a central nucleus, is very interesting. But I shall never be content, for one, until I can see such an atomic system in operation.'

"Now I had every admiration for the capacity and genius of my teacher, but I couldn't forebear exclaiming:

"'Is that possible?'

"'Of course it's possible,' he cried irritably. 'Do you think I should be pursuing my experiments if I didn't think it possible? Only numbskulls think anything impossible!'

"I FELT rather hurt at his retort and a certain coolness sprang up between us. The summer holidays came and I went away without bidding him good-by. But returning for the new semester, my first act was to hurry to the laboratory. He greeted me as if there had never been any difference between us.

"'Come,' he cried; 'you must see what I have accomplished. It is marvelous, marvelous.'

"In his workshop stood a mechanism perhaps three feet square and four feet high. It was made of polished steel and looked not unlike an Edison music box.

"'You are the first I have shown it to,' he said excitedly. 'Here, look into this.'

"Stooping over the top of the box I peered into the eye-piece indicated. It was so fashioned that it fitted the contour of the face snugly.

"'Now hold steady,' warned the Professor. 'This machine makes quite a noise, but it won't harm you at all.'

"I sensed that he was fingering and arranging dials and levers on the side of the contrivance. Suddenly an engine in the box began to throb with a steady rhythm. This gradually increased in tempo until the vibration of it shook the room.

"'Don't move,' shouted the Professor.

"At first I could see nothing. Everything was intensely dark. Then the darkness began to clarify. Or rather I should say it seemed as if the darkness increased to such a pitch that it became—oh, I can't describe it! But of a sudden I had the sensation of looking into the utter bleakness and desolation of interstellar space. Coldness, emptiness—that was the feeling. And in this coldness and emptiness flamed a distant sun, around which twelve darker bodies the size of peas revolved. They revolved in various ellipses. And far off—millions of light years away (the thought came to me involuntarily at the time)—I could glimpse infinitesimal specks of light, a myriad of them. With a cry I jerked back my head.

"'That,' shouted the Professor in my ear, 'was an atomic universe.'

"IT never entered my head to doubt him. The realness, the vividness, the overwhelming loneliness and vastness of the sight I had seen—yes, and the suggestion of cosmic grandeur and aloofness that was conveyed—banished any other feeling but that of belief.

"'Inside that box,' said Professor Reubens quietly, 'and directly underneath the special crystal-ray medium I have perfected, is a piece of matter no larger than a pin-head. But viewed through the magnifying medium of the crystal-ray that insignificant piece of matter becomes as vast and as empty as all space, and in that space you saw—an atomic system.'

"An atomic system! Imagine my emotions. The tremendousness of the assertion took away my breath. I could only seize the Professor's hand and hold to it tightly.

"'Softly, my boy, softly,' he said, smiling at my emotion. 'What you have seen is but the least part of the invention. There is more to it than that.'

"'More?'

"'Yes. Did you think I would be content with merely viewing at a distance? No. Consider that revolving round a central nucleus similar to our sun are twelve planets, any one of which may be inhabited by intelligent creatures.'

"I stared at him dumbly.

"'You mean—'

"'Why not? Size is only relative. Besides in this case I can demonstrate. Please look again.'

"Not without trepidation, I did as he bade. Once more I saw the black emptiness of atomic space, saw the blazing nucleus with its whirling satellites. Above the roaring noise of the machine came Professor Reubens' voice. 'I am now intensifying the magnifying medium and focusing it on one of the planets you see. The magnifying crystal-ray is mounted on a revolving device which follows this particular planet in its orbit. Now ... now....'

"I GAZED, enthralled. Only one atomic planet—the size of a pea and seemingly motionless in space—now lay in my field of vision. And this planet began to grow, to expand, until beneath my staring eyes it looked like the full moon in all its glory.

"'I am gradually increasing the magnifying power of the crystal-ray,' came the voice of the Professor.

"The huge mass of the planet filled the sub-atomic sky. My hands gripped the rim of the box with excitement. On its surface began to form continents, seas. Good God! was all this really materializing from a speck of matter under the lens of a super-microscope? I was looking down from an immense height upon an ever clarifying panorama. Mountains began to unfold, plains, and suddenly beneath me appeared a mighty city. I was too far away to see it distinctly, but it was no city such as we have on earth. And yet it was magnificent; it was like gazing at a strange civilization.

"Dimly I could see great machines laboring and sending forth glowing streamers of light. Strange buildings rose. It was all bizarre, bewildering, unbelievably weird. What creatures dwelt in this place? I strained my eyes, strove to press forward, and in that very moment the things at which I gazed seemed to rise swiftly to meet my descending head. The illusion was that of plunging earthward at breakneck speed. With a stifled cry, I recoiled, rubbed my blinking eyes, and found myself staring stupidly into the face of Professor Reubens. He shut off the machine and regarded me thoughtfully.

"'In that atomic universe, on a planet swinging round a sub-atomic sun, the all of which lies somewhere in a speck of our matter, intelligent creatures dwell and have created a great machine civilization. And Baxter,' he leaned forward and fixed me with eyes that gleamed from under heavy brows, 'not only has my super-atomic-microscope revealed somewhat of that world and its marvels to human vision, but it has opened up another, a more wonderful possibility.'

"HE did not tell me what this wonderful possibility was, and a few minutes later I left the laboratory, intending to return after a late class. But a telegram from Phoenix was at my rooms, calling me home. My father was seriously ill. It was June before he recovered his health. Consequently I had to forego college until the next season.

"'Old Reubens is going dotty,' said one of my classmates to me. Rather disturbed, I sought him out. I saw that there were dark circles of sleeplessness under his eyes and that his face had grown thinner. Somewhat diffidently I questioned him about his experiments. He answered slowly:

"'You will recollect my telling you that the super-atomic-microscope had opened up another wonderful possibility?'

"I nodded, sharply curious now.

"'Look.'

"He led the way into his workshop. The super-atomic-microscope, I noticed, had been altered almost out of recognition. It is hopeless for me to attempt describing those changes, but midway along one side of its length projected a flat surface like a desk, with a large funnel-shaped device resting on it. The big end of this funnel pointed towards a square screen set against the wall, a curious screen superimposed on what appeared to be a background of frosted glass.

"'This,' said the Professor, laying one hand on the funnel and indicating the screen with the other, 'is part of the arrangement with which I have established communication with the world in the atom.

"'No,' he said, rightly interpreting my exclamation, 'I am not crazy. For months I have been exchanging messages with the inhabitants of that world. You know the wave and corpuscular theories of light? Both are correct, but in a higher synthesis—But I won't go into that. Suffice it to say that I broke through the seemingly insuperable barrier hemming in the atomic world and made myself known. But I see that you still doubt my assertion. Very well, I will give you a demonstration. Keep your eyes on the screen—so——'

"ADJUSTING what seemed a radio headpiece to my ears, he seated himself at a complicated control-board. Motors purred, lights flashed, every filament of the screen became alive with strange fires. The frosted glass melted into an infinity of rose-colored distance. Far off, in the exact center of this rosy distance appeared a black spot. Despite the headpiece, I could hear the Professor talking to himself, manipulating dials and levers. The black spot grew, it advanced, it took on form and substance; and then I stared, I gasped, for suddenly I was gazing into a vast laboratory, but depicted on a miniature scale.

"But it wasn't this laboratory which riveted my attention. No. It was the unexpected creature that perched in the midst of it and seemed to look into my face with unwinking eyes of gold set in a flat reptilian head. This creature moved; its feathers gleamed metallically; I saw its bill open and shut. Distinctly through the ear-phones came a harsh sound, a sound I can only describe by the words toc-toc, toc-toc. Then, just as the picture had appeared, it faded, the lights went out, the purring of the motors ceased.

"'Yes,' said the Professor, stepping to my side and removing the headpiece, 'the inhabitants of the sub-atomic planet are birds.'

"I could only stare at him dumbly.

"'I see that astounds you. You are thinking that they lack hands and other characteristics of the genus homo. But perhaps certain faculties of manipulation take their place. At any rate those birds are intelligent beings; in some respects, further advanced in science than are we ourselves. Perhaps it would be more exact to say that their scientific investigations and achievements have been along slightly different lines. If such messages I sent them had come to our world from another planet or dimension, how readily they might have been misconstrued, ridiculed or ignored.' The Professor shrugged his shoulders. 'But the beings in this sub-atomic world interpreted my communications without difficulty.

“IN no time we were conversing with one another through means of a simplified code. I was soon given to understand that their scientists and philosophers had long recognized the fact that their universe was but an atom in an immeasurably greater dimension of existence; yes, and had long been trying to establish contact with it.' The Professor's voice fell. 'And not that alone: they were eager to cooperate with me in perfecting a method of passing from their world to ours!

"'Yes,' he cried, 'much of what I have accomplished has been under their advice and guidance; and they on their part have labored; until now'—his eyes suddenly blazed into my fascinated face—'until now, after months of intensive work and experiment, success is nigh, and any day may see the door opened and one of them come through!'

"Gentlemen!" cried Milton Baxter, "what more is there to say? I staggered from Professor Reubens' laboratory that afternoon, my head in a whirl. That was on a Monday.

"'Come back Thursday,' he said.

"But as you know, Professor Reubens disappeared on a Wednesday night before; and stranger still, his machines disappeared with him. In his laboratory were signs of a struggle, and bloodstains were found. The police suspected me of a guilty knowledge of his whereabouts, in short of having made away with my friend. When I told somewhat of the experiments he had been engaged in, spoke of the missing inventions, they thought I was lying. Horrified at the suspicion leveled at myself, I finally left Tucson and went abroad. Months passed; and during all those months I pondered the mystery of the Professor's fate, and the fate of his machines. But my fevered brain could offer no solution until I read of what was happening in Arizona; then, then...."

Milton Baxter leaned forward, his voice broke.

"Then," he cried, "then I understood! Professor Reubens had succeeded in his last experiment. He had opened the door to earth for the bird intelligences from the atom and they had come through and slain him and spirited away his machines and established them in a secret place!

"God help us," cried Milton Baxter, "there can be but one conclusion to draw. They are waging war against us with their own hideous methods of warfare; they have set out to conquer earth!"

SUCH was the amazing story Milton Baxter told the Senate, but that body placed little credence in it. In times of stress and disaster cranks and men of vivid imaginations and little mental stability inevitably spring up. But the Washington correspondents wired the story to their papers and the Associated Press broadcast it to the four winds.

Talbot had just returned to Phoenix from New Mexico. He had been out of touch with civilization and newspapers and it was with a feeling of stunned amazement that he learned of the evacuation of Tucson and Winkleman and the wiping out of Oracle. Reading Milton Baxter's incredible story he leapt to his feet with an oath. Toc-toc! Why, that was the sound the strange birds had uttered in the hills back of Oracle. And there was the noise of machinery coming from the old shaft.

Full of excitement he lost no time in seeking an interview with the military commander whose headquarters were located in Phoenix and related to him what Manuel and himself had witnessed and heard that day at the abandoned mine. Manuel corroborated his tale. The commander was more than troubled and doubtful.

"God knows we cannot afford to pass up an opportunity of wiping out the enemy. If you will indicate on a map where the old shaft is we will bomb it from the air."

But Talbot shook his head.

"Your planes would have a tough job hitting a spot as small as that from the air. Besides, a direct hit might only close up the shaft and not destroy the workings underground. If the enemy be the creatures Milton Baxter says they are, what is to prevent them from digging their way out and resuming the attack?"

"Then we will land troops in there somehow and overwhelm them with——"

TALBOT interrupted. "Pardon me, General, but the enemy would have no difficulty in spotting such a maneuver. What chance would your soldiers have against a shower of jungle seed? You would only be sending them to destruction. No, the only way is for someone familiar with those old underground diggings to enter them, locate the birds and the machines and blow them up."

"But who— —"

"Myself. Listen. This is the plan. About five years ago my company mined for copper and other ores about a half mile above the Wiley claim. I was in charge of operations. That is how I know the ground so well. One of our northern leads broke through into a tunnel of the abandoned mine. When copper prices were shot to hell in the depression of 1930 we quit taking out ore; but when I went through the place eighteen months ago it was still possible to crawl from one mine to another. Of course earth and rock may have fallen since then, but I don't believe the way is yet blocked. If I were dropped in that vicinity at night with another man and the necessary tools and explosives...."

The general thought swiftly.

"An auto-gyroscope could land you all right. There's one here now. But what about the second man to accompany you?"

Manuel said quickly, "I'm going with the boss."

"You, Manuel," Talbot said roughly. "Don't be a fool. If anything should happen to me—well, I've lived my life; but you're only a kid."

Manuel's face set stubbornly. "An experienced mining man you need, is it not? In case there should be difficulties. And I am experienced. Besides, señores," he said simply, "my wife and child are somewhere in those mountains ... above Oracle...."

Talbot gripped his hand in quick sympathy. "All right, Manuel; come if you like."

A MOONLESS sky hung above them as they swung over the dark and jungle-engulfed deserted city of Tucson, a sky blazing with the clarity of desert stars, and to the south and west shot through with the beams of great searchlights. Flying at a lofty altitude to avoid contact with drifting globes or betrayal of their coming with no lights showing aboard their craft save those carefully screened and focused on the instrument board, it was hard to realize that the fate of America, perhaps of the world, hung on the efforts of two puny individuals.

Everything seemed unreal, ghost-like, and suddenly the strangeness of it all came over Talbot and he felt afraid. The noiseless engine made scarcely a sound; the distant rumble of gunfire sounded like low and muttering thunder. They had come by way of Tucson so as to pick up a ten-gallon tube of concentrated explosive gas at the military camp in the Tucson mountains.

"This gas," the general had assured them, "has been secretly developed by the chemical branch of the War Department and is more powerful than TNT or nitro-glycerin. It is odorless, harmless to breathe and exploded by a wireless-radio device."

He had showed them how to manipulate the radio device, and explained that in the metal tube was a tiny chamber from which gas could not escape, and a receiving-detonating cap. "If you can introduce the tube into the underground galleries where you suspect the enemy's headquarters to be, allow the contents to escape for ten minutes, and a mile distant you can blow the mine and all in it to destruction. And you needn't be afraid of anything escaping alive," he had added grimly.

TALBOT thought of his words as the dark and silent world slid by. He glanced at the luminous dial of his wrist-watch. Eleven-fifteen. The moon rose at eleven-twenty-four. He studied the map. High over Mount Lemmon the craft soared. He touched the army pilot's arm. "All right," he said, "throttle her down." Their speed decreased. "Lower."

Swiftly they sank, until the dark bulk of hills and trees lay blackly beneath; so near as to seem within the touch of a hand. Though he strained his ears, no alien sound came wafting upward. "Keep circling here," he directed the pilot. "The moon'll be up in a minute and then we can be sure of where we are." The pilot nodded. He was a phlegmatic young man. Not once during the trip had he uttered a word.

The east glowed as if with red fire. Many a time before had Talbot watched the moon rise, but never under stranger circumstances. Now the night was illuminated with mellow glory. "Hit the nail on the head," he whispered. "Do you see that spot over there? To the left, yes. Can you land us there?"

Without a word the pilot swung for the clearance. It was a close thing, requiring delicate maneuvering, and only an auto-gyroscope could have made it without crashing. Hurriedly Manuel and Talbot unloaded their gear.

"All right," said Talbot to the pilot. "No need to wait for us. If we are successful, we'll send out the wireless signal agreed on, and if we aren't...." He shrugged his shoulders. "But tell the General to be sure and allow us the time stipulated on before undertaking another attack."

STANDING there on the bleak hillside, watching the auto-gyroscope run ahead for a few yards and then take the air, Talbot experienced a feeling of desolation. Now he and Manuel were alone, cut off from their own kind by barriers of impregnable jungle. And yet on that lonely hillside there were no signs of an enemy. For a moment he wondered if he weren't asleep, dreaming; if he wouldn't soon awake to find that all this was nothing but a nightmare.

But Manuel gathering up the tools aroused him from such thoughts. Not without difficulty were the necessary things conveyed to the abandoned mine back of the old Wiley claim. Their course lay along the bottom of a dry creek, over a ridge, and so to the shaft half-way down the side of a hill. A second trip had to be made to bring the gas tube.

It was two o'clock in the morning when Manuel stood at the foot of the four-hundred-foot hole and signaled up that the air was good. Talbot lowered the tools to him, and the gas container, and lastly went down himself. As already stated, Talbot had explored the underground workings of the mine not eighteen months before. Picking out the main tunnel and keeping a close watch for rattlers with electric torches, the two men went cautiously ahead. In places earth had fallen and had to be cleared away, but the formation for the most part was a soft rock and shale. They went slowly, for fear of starting slides.

At a spot taking an abrupt turn—and it was here that the newer tunnel had broken through into the older gallery of the Wiley claim—Manuel caught swiftly at Talbot's arm. "What is that?" To straining ears came the unmistakable throb of machinery. They snapped off their torches and crouched in Stygian darkness. Not a ray of light was to be seen. Talbot knew that in following the ore stratum, the Wiley gallery took several twists. Laboriously he and Manuel advanced with the gas tube. It was stiflingly close. He counted the turns, one, two, three. Now the roar of machinery was a steady reverberation that shook the tunnel. He whispered to Manuel:

"Go back and wait for me at the mouth of the shaft. Only one of us must risk taking the gas tube any nearer the enemy. Here, take my watch. It is now two-forty-five. If I don't rejoin you by four o'clock touch off the explosive."

Manuel started to protest. "Do as I say," commanded Talbot. "The fate of the world is at stake. Give me an hour; but no longer—remember!"

LEFT alone in the clammy darkness Talbot wiped the sweat from his face. Grabbing one end of the rope sling in which the tube was fastened, he pulled it ahead. There was a certain amount of unavoidable noise; rock rattled, earth fell; but he reasoned shrewdly enough that the roar of the machinery would drown this. Beyond a crevice created by a cave-in he saw an intense light play weirdly. He squirmed through the crevice and pulled the tube after him.

His mind reconstructed the mine ahead. He recollected that when the lead of this mine had petered out, the owners had begun to sink the shaft deeper into the earth before abandoning the mine. This meant that the foot of the shaft, with the addition of an encroaching twenty feet of the southern gallery, was deeper by some several yards than the floor of the tunnel in which he stood. Here was the logical place to set the gas tube, nose pointed ahead.

With trembling fingers he loosened the screwed-in nose of the tube with a wrench. A slight hiss told of the deadly gas's escape. It would inevitably flow towards the shaft, drawn by the slight suction of machinery, following the easiest direction of expansion. Now Talbot's work was done, and if he had immediately retreated all would have been well, but the weird light fascinated him. Here he was, one man in the bowels of earth pitting his strength, his ingenuity against something incredible, unbelievable. Beings from an atomic universe, from a world buried within the atom; beings attacking his own earth with uncanny methods of destruction. Oh, it was impossible, absurd, but he must look at them, he must see.

Scarcely daring to breathe, he squirmed, he crawled, and suddenly he saw. He was looking down into an underground crypt flooded with brilliant light. That crypt had been altered out of all recognition, its greater expanse of roof supported with massive pillars, the light screened away from the shaft. But it was not all this which riveted his staring eyes. No—it was the machines; strange, twisted things, glowing, pulsing, and—in the light of his knowledge—menacing and sinister.

TALBOT gasped. Almost at once he observed the birds, twelve of them, two standing in front of what appeared to be a great square of polished crystal, wearing metal caps and goggles, heads cocked forward intently. The others also perched in front of odd machines like graven images. That was the uncanny thing about the birds: they appeared to be doing nothing. Only the occasional jerk of a head, the filming of a hard golden eye, gave them a semblance of life. But, none the less, there could be no mistaking the fact that they were the guiding, the directing geniuses back of all the pulsing, throbbing mechanisms.

Half mesmerized by the sight, forgetful of time and place, Talbot leaned forward in awe. There was a great funnel, a shallow cabinet, and out of the cabinet poured an intense reddish beam, and out of the beam....

It was a minute before he understood, and then comprehension came to him. Those dark spots shooting from the cabinet, no larger than peas, were the mysterious drifting globes whose scattered seed was fast covering miles of Arizonian soil with impenetrable jungle. From a universe in a piece of matter no larger than a pin-head, from a sub-atomic world, the weapons of an alien intelligence were ruthlessly being hurled against man, to conquer, to destroy him.

And now it was made plain to him why the drifting globes had seemed to materialize out of thin air. Being infinitesimally small parts of an atom, these globes were released from the cabinet and soon assumed the size of peas; they were guided across the crypt, up the old Wiley shaft, and high in the air, somewhere in space, enlarged to immense proportions. How? Talbot could not guess. By some manipulation of science and machinery beyond that of earth.

Engrossed, he moved an inch forward, craned his head, and in that moment it happened. Beneath his weight a section of earth and rock crumbled, cracked, slid forward, and he plunged headlong to the floor below, striking his skull with stunning force!

HE came to himself, staring up into the dour-looking face of a tall man. He recollected pitching forward among the birds and the machines. But the birds and the machines had disappeared and he was lying in an odd room without windows but lit with a soft radiance. Bewildered, he sat up.

"Who are you?" he demanded.

The man's beard looked straggly, untrimmed.

"My name," he said, "is Reubens—Professor Reubens."

Professor Reubens! Talbot gasped. "Not the scientist who disappeared?"

"Yes—as you've disappeared."

"What!"

"Through the machine."

It was a moment before Talbot understood. "You mean...."

"That you are a prisoner in a sub-atomic world."

Talbot now realized with startling clearness what had happened to him. When he had fallen into the crypt the weird birds had directly placed him in the cabinet and transported him to their own world. In other words, he and Reubens and everything he saw about him were infinitely small creatures in an atom-world. He and the Professor were trapped! And when Manuel blew up the only means of return....

"How long have I been here?" Talbot asked hoarsely.

"Five minutes at the most."

Then, at the shortest, the way to earth would exist twenty minutes longer. Twenty minutes.... Incoherently he told Reubens of what had happened in Arizona since his disappearance, of his own misadventure.

"Aye," said the Professor, "I knew as much. Nor do these inhuman birds intend stopping with the use of seed globes. More devilish weapons than that they plan using against earth. Oh, they are fiends, fiends! Already have they wiped out civilization and intelligent life on other planets in this sub-atomic system and introduced their own."

HE stopped, shuddering. "Nor is it to be wondered at that no birds were seen after the first attack on Oracle," he went on. "They do not fight in person, as do we ourselves, but through proxy, directing machines from centers of control. In powers of destruction, they are immeasurably ahead of man. Thank God you discovered their headquarters in the deserted mine and have spread the gas for its destruction. But the rage of the birds at such a defeat will be terrible. They will undoubtedly torture me in an effort to make me reveal the basis of my invention so that they can resume the attack on earth. So we must escape."

"But how—where?"

"I have thought that out. It is one chance in a thousand. Undoubtedly we will be killed. But that is better than being tortured or living in this world. Look."

He held up a pearl-handled pen-knife. "The birds are smart, all right, but they don't quite understand clothes, wearing none themselves. They found your revolver, but overlooked this."

"Of what good is it?"

"To cut our way out of this cell."

Talbot laughed incredulously. The walls of the room were smooth, and hard to the touch. "They're as solid as concrete," he said.

"But cut like cheese under a steel blade. I found that out. Watch."

To Talbot's amazement the point of the penknife sank into the wall and in a moment a section of it was gouged out. The professor said tensely, "I've been months in this place, been taken back and forth, and know the lay of the land. This room is in a great building that houses the laboratory from which the attack against earth is being launched. Would you believe it, only the great scientist who picked up my messages and helped me perfect my invention, and a few of his assistants, are concerned in that attack, and they will be congregated at the machines. Follow me, and whatever I command, do it promptly."

THE Professor had been working feverishly as he spoke, and now he and Talbot crawled through the hole he had made in the wall and found themselves in a long gloomy corridor. "Quick," Reubens whispered.

They darted down the passageway. Talbot had only time to see that the gleaming sides of the corridor were beveled and etched with strange designs, before they came to its end and where a curious device like a huge five-pointed star was revolving noiselessly, half sunk in a great hole in the floor. Without hesitation the Professor stepped onto one of the flat-tipped star-points as it came level with where they stood and Talbot did the same. Up, turned the star-point, to a dizzy height, and over, but the tip swung on ball-bearings, maintaining its passengers in a perpendicular position, and from its highest point of elevation descended to another floor far below, where they disembarked.

The huge revolving star-wheel was nothing but an ingenious movable staircase. But the Professor gave Talbot no time to marvel, nor did the latter try to linger. The corridor below was wider, more richly beveled and carved, and the statue of an heroic bird stood perched in the center of it. The lighting was soft and mellow, but Talbot could perceive no windows or globes. Suddenly from an open doorway hopped a bird. There was no chance to avoid it. Its wings were spread and from its parted bill came a harsh cry, "Toc-toc, toc-toc!"

KNIFE in one hand, the Professor hurled himself forward and caught the bird in the grip of the other. Instantly from the doorway sprang a monstrous mechanism on stilts, flexible tentacles of metal reaching out and wrapping themselves around the Professor. Talbot leaped to the Professor's assistance. The mechanism fought like a live thing. In vain he strove to wrench the tentacles free of the Professor. One of them lashed out and took him by the thighs in a crushing grasp. But the Professor had the bird by the throat. Both of his hands were free. Back, he forced its head, back. The mechanism seemed to falter in the attack, as if bewildered. Across the exposed throat the Professor drew the gleaming blade. Flesh, tendons and arteries gave, blood spurted, and in the same moment the tentacles fell away from Talbot and the Professor and withdrew with a dull clang. The Professor released the bird and it dropped to the floor.

"It is the birds' mentality that directs those mechanisms," said the Professor, pointing to the now harmless machine.

Apparently the brief but terrific battle had passed unnoticed, no alarm being given. Now the corridor twisted. The two men came to where a deep well was sunk in the floor. To one side a star-wheel revolved smoothly. Out of the depths came the steady throb of machinery. Cautiously peering over the edge, Talbot saw a sight he would never forget.

HE did not need the Professor's whispered words to tell him that here was the source of the deadly attack being waged against earth. Motionless birds perched in front of bizarre machines; lights waxed and waned; a cannon-like device, or funnel, shot a column of light into a screen, and through the column of light moved a steady procession of round objects the size of plums.

"The drifting globes being shot through to earth," whispered the Professor, "and our only hope. Listen, the birds are intent on their machines, their backs to the star-wheel. We will descend, throw ourselves into the column of light, seize hold of a globe, and...."

He did not need to finish. Talbot understood in a flash. They would be dragged to their own world by the weapons hurled at it.

"Of course that column of light may kill us," went on the Professor tensely. "Or we may be blown up on the other side. Your Mexican friend hasn't touched off that explosive gas yet, because—But we've not a moment to lose. Follow me."

The tip of the star-wheel went up, over, descended. The blood was roaring in Talbot's ears. "Now!" hissed the Professor. "Now!" Together they rushed forward. Talbot's foot slipped. The heart leaped into his throat. He never remembered reaching the column of light; but suddenly he was in it, blinded, dazed. His clutching hands closed on something small and hard.

The laboratory was a pinwheel going round and round. Through a sea of darkness he floated. A distant glow grew, expanded, became the crypt in the old Wiley mine. A moment he glimpsed the gleaming pillars, the pulsing machines, the startled birds, and then—Oh, it was incredible, impossible, but the dark, crumbling walls of the old shaft were around him; the globe in his hand no larger than a pea was lifting him towards life and safety.

He wanted to shout, to sing, but even as the pale stars fell athwart his upturned face, even as the cool mountain air smote his fevered brow, the dark earth erupted beneath his feet, a whirlwind of smoke and wind beat and buffeted him, and, in the midst of an overwhelming noise, consciousness was blotted out!

It was bright daylight when Talbot regained his senses. Propped against a great rock the Professor regarded him whimsically. Reubens looked badly bruised and battered; one arm hung loosely at his side. Talbot's head ached and he knew that a leg was broken.

"Yes," said the Professor, "we got through just in time—a few seconds before the explosive gas was touched off. Thank God, my invention has been destroyed. The world is safe."

Yes, the world was safe. Talbot sank back with a sigh of relief. Overhead a white plane was dipping toward earth.